Work and Machines

Printed in Mexico

ISBN-13: 978-0-15-362036-2

ISBN-10: 0-15-362036-6

2 3 4 5 6 7 8 9 10 805 16 15 14 13 12 11 10 09 08

SCHOOL PUBLISHERS

Visit *The Learning Site!*
www.harcourtschool.com

VOCABULARY
work

What Is Work?

Work is done when a push or pull is used to move something. This girl is doing work. She is pulling a wagon.

READING FOCUS SKILL
MAIN IDEA AND DETAILS

The **main idea** is what the text is mostly about. **Details** tell more about the **main idea**. Look for details about work.

Work

Is doing a math problem work? A scientist would say no. Is kicking a ball work? A scientist would say yes.

To a scientist, **work** is done only when a force is used to move something through a distance. A *force* is a push or a pull.

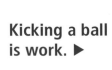

Kicking a ball is work. ▶

4

Throwing a ball is work, because you use force to move an object. Your hand pushes the ball. This makes the ball move.

 Tell how the children in the picture are doing work.

Play can be work. ▼

Measuring Work

You use different amounts of force to do work. To move a heavy box, you use a large amount of force. To move a light box, you use less force. Which box would take more force to move?

Box is full.

Box is half full.

How can you tell how much work it takes to move an object? You need to measure two things. You need to measure how much force you use. You also need to measure how far the object moves.

Focus Skill **What two measurements tell how much work is done?**

◀ **The boy is using force to move the girl on the board.**

Review

Complete this main idea statement.

1. _____ is done when a force is used to move something.

Complete these detail sentences.

2. A force is a push or a _____.

3. You can _____ how much work it takes to move an object.

What Are Some Simple Machines?

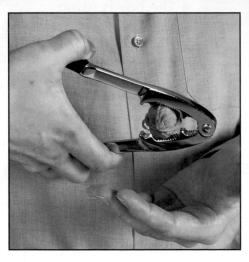

fulcrum

A **simple machine** has few or no moving parts. It helps you do work. This nutcracker is a simple machine.

A **lever** is a bar that turns on a fixed point. The fixed point is called the **fulcrum**. A broom and a rake are levers.

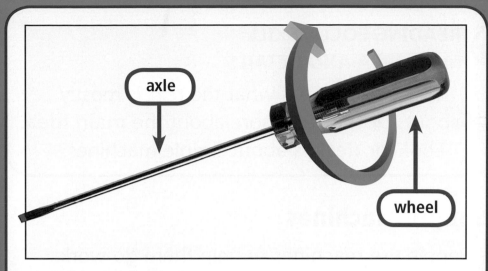

A **wheel-and-axle** is a simple machine that has a wheel and an axle that turn together. A screwdriver can be a wheel-and-axle.

A **pulley** is a wheel with a rope around it. A pulley is a simple machine that makes it easier to lift things.

READING FOCUS SKILL
MAIN IDEA AND DETAILS

The **main idea** is what the text is mostly about. **Details** tell more about the **main idea**. Look for **details** about simple machines.

Simple Machines

People use machines to help them do work. Some machines, such as cars, have many parts that move. Other machines, like a ramp, have no parts that move. **Simple machines** have few or no moving parts.

 Tell why a broom is a simple machine.

A broom is a simple machine. ▶

fulcrum

◀ A rake is a lever.

The Lever

One kind of simple machine is a lever. A **lever** is a bar that turns on a point that does not move. The point is called a **fulcrum**. Levers make work easier.

A rake is a kind of lever. It makes gathering leaves easier. The hand that holds the end of a rake does not move. This is the fulcrum. The other hand moves the middle of the handle. This makes the end of the rake move even more and gather more leaves.

 Tell why a rake is a lever.

The Wheel-and-Axle

A **wheel-and-axle** is a simple machine. It has a wheel and an axle joined together. When you turn the wheel, the axle also turns.

A screwdriver is a wheel-and-axle. The handle is the wheel. The rod is the axle. When you turn the handle, the rod also turns. The handle makes it easier to turn the rod. It lets you use less force. This makes work easier.

Focus Skill **Tell how a wheel-and-axle makes work easier.**

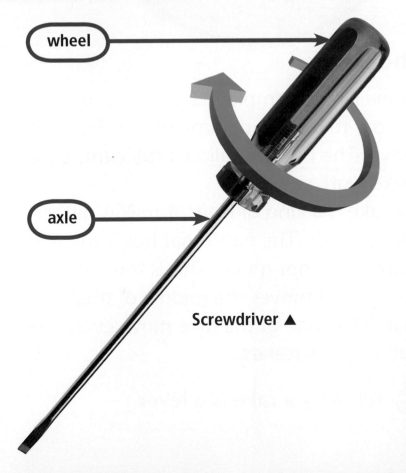

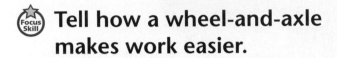

wheel

axle

Screwdriver ▲

The Pulley

A **pulley** is a wheel with a rope around it. It is used to lift things. When you pull down on one end of the rope, the pulley makes the other end move up.

Flagpoles use pulleys. When you pull down on the rope, the flag moves up the pole.

 What is a pulley used for?

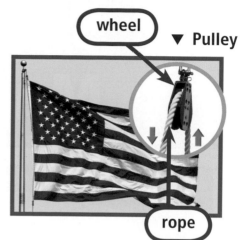

wheel

▼ Pulley

rope

 Complete this main idea statement.

1. A _____ has few or no moving parts.

Complete these detail statements.

2. A broom and rake are kinds of _____.

3. A wheel-and-axle turn _____ on a screwdriver.

4. A _____ makes it easier to lift something.

VOCABULARY

inclined plane
wedge
screw

What Are Some Other Simple Machines?

An **inclined plane** is a simple machine. It is a slanted surface that makes it easier to move and lift things. A ramp is a kind of inclined plane.

A **wedge** is a simple machine that can split things. It is made from two inclined planes. An ax is a wedge.

A **screw** is a simple machine you turn. A screw can hold things together. It can also lift things.

When you **compare and contrast** you tell how things are alike and different.

Look for ways that inclined planes, wedges, and screws are alike and different.

Inclined Planes

The path up this hill is an inclined plane. An **inclined plane** is a slanted surface that makes it easier to move or lift things. It is a kind of simple machine.

The path is an
inclined plane. ▶

Ramp ▲

A ramp is also an inclined plane. A ramp makes it easier to move and lift things. Many people push things up ramps to get them onto trucks.

 **How is a ramp like a path up a hill?**

This boy is using an inclined plane. ▶

The Wedge

A wedge is a kind of simple machine. A **wedge** is two inclined planes put back-to-back. When you press down on a wedge, it splits things in two. A knife is a wedge. An ax is also a wedge. It is used to split logs.

 Tell how a wedge and an inclined plane are alike and different.

▼ A knife is a wedge.

▼ An ax is a wedge.

Screw

A **screw** is a simple machine people turn to lift an object. It is also used to hold two or more objects together, such as pieces of wood.

A screw looks like a nail with threads wrapped around it. The threads help keep the screw in objects. They also help hold objects together.

 Tell how a screw and an inclined plane are alike.

Screws hold wood together. ▶

Review

 Complete these sentences to compare and contrast simple machines.

1. An inclined plane, a wedge, and a screw are all _____.

2. A _____ is made up of two inclined planes put back-to-back.

3. Both an inclined plane and a _____ make lifting easier.

GLOSSARY

fulcrum (FUHL•kruhm) The fixed point on a lever (11)

inclined plane (in•KLYND PLAYN) A slanted surface that makes moving or lifting things easier (16)

lever (LEV•er) A simple machine made up of a bar that pivots, or turns, on a fixed point (11)

pulley (PUHL•ee) A simple machine made up of a wheel with a rope around it (13)

screw (SKROO) A simple machine that you turn to lift an object or to hold two or more objects together (19)

simple machine (SIM•puhl muh•SHEEN) A tool with few or no moving parts that helps people do work (10)

wedge (WEJ) A simple machine that is made up of two inclined planes placed back-to-back (18)

wheel-and-axle (weel•and•AK•suhl) A simple machine made up of an axle connected to a wheel that both turn together (12)

work (WERK) The use of a force to move an object (4)